The MOST ENERGETIC SOURCE OF ENERGY

by

G. P. W a l i a

Printed and published in USA, by

Xlibris Corporation
1663 Liberty Drive, #200,
Bloomington, IN 47403 USA

Phone: 888-795-4274
Fax: 610.915.0294 / 610.915.0295
e-mail: Orders@Xlibris.com
        www.Xlibris.com

Library of Congress Control Number: 2011912796
ISBN: Softcover   978-1-4653-3984-3
                  Ebook   978-1-4653-3985-0

This book is available at most of the leading book stores all over the world as well as at major retail outlets on the internet, including *amazon.com, barnesandnobles.com, etc.etc.*

Price U.S.$15.99
plus Shipping & Handling chargers as applicable.

On bulk purchase of this book, special discounts are available. For details, please contact the publisher.

<u>E N D E A V O R</u>

a tiny step forward

to address

the most burning issues

of our present times,

the **ENERGY**

&

the **POLLUTION**

(Author)

BASIS

Entire known history

of the entire mankind

is the history of

only a few individuals

whose dreams and actions

were different

from the whole lot.

(Swami Vivekananda)

Dear __READERS__

Like many other authors, I do wish to populate
several such pages with lots of degrees, diplomas,
certificates, prizes, awards, rewards etc. etc. and etc.

Since the inception of my carried-over memories,
I have always been identifying myself as one of the
most studious students, aspiring to master all the
sciences of life.

Several decades have gone bye and I am still a
student at the lower primary level. Quite often, my
in-abilities have aggravated my in-capabilities and
kept the ultimate results uniformly deficient.

In spite of countless set-backs and failures, I consider
my time and efforts that have gone into my endeavors
to learn, have been amply rewarding.

I am learning to notice, to observe, to analyze and to
evaluate the entire history and geography of my own
self as well as of those around me. And, in that light,
I am learning to decide almost most of my actions,
my re-actions and also my in-actions.

I am learning to get access to the infinite galaxy of
thinking and fully awake dreams. Also I am learning
to choose my dreams, grade them, pack them neatly
and treasure them safely in the most appropriate
portions of my heart, my psyche and my memory.

I am learning to invite the entire mankind to freely share the entire treasure of my knowledge.

It makes me feel closely related to the whole world; the world without borders, partitions, divisions or distinctions whatsoever.

I am learning to learn all this as best as possible. I do hope to go on learning much more.

With lots of hopes for very bright horizons for YOU, for ME and for ALL OF US.

G. P. Walia

The <u>QUEST</u>

PERFECTION
in any sphere
remains so
only for
the shortest measurable
fraction of time.

As the world turns
the realities change
and
so change the demands of realities
leaving behind
an infinite trail
of imperfect perfections
which trigger
the un-ending quest
for more perfections.

C O N T E N T S

PRESENTATION   1   >   8

REALITIES   9   >   12

IN-SIGHT   13   >   22

SUGGESTIONS   23   >   38

OPPORTUNITIES   39   >   50

GO-AHEAD   51

Notes & Remarks   52   >   54

REALITIES

NECESSITY is the mother of invention. And, human brain is always the biological father. Since inception, this growing-growing-grown process is being called the "Human Evolution".

With the super fast pace of industrialization, overall demand for energy is already riding the rockets. At the same time, most of the known natural treasures of conventional sources of energy are already showing their bottoms. Unless and until any dependable alternate source of energy is available, the future of this glamorous world can become tragically dark.

Around the clock, all over the world, the best known wizards and experts are wrestling with the possible options and dependable solutions. Over the time, it has given birth to several short lived instant solutions as well as a few longer lasting ones. But, any ideal solution is still evading.

Ultimate choice of the ideal source of energy would bank upon its being freely and abundantly available at all the time and at all the places, all over the world. Its procurement and utilization should be technically feasible and economically viable. The utilized stocks could preferably be 'self-replenishing' or at least be readily replenish-able. It must be more environment friendly and pollution free than all other sources.

Final choice of the preferred sources of energy would determine the future shapes, sizes and complexions of various production establishments and the structure of their demands for specific industrial inputs.

Conclusive options would obviously depend upon the overall commercial and financial interests at stake through the switch-over to any suggested alternate source of energy.

Alternate sources of energy being suggested these days include the Rectified Coal or Clean Coal as they call it. Some ultra-processed Fossil Fuels are also being pushed to the front rows. Some Bio Fuels are being promoted by a number of giants in the field. Also in the field are, the Nuclear Energy, Geothermal Energy, Solar Energy, Natural Gas and GTL, Tidal and Hydro Power, Hydrogen Fuel Cells and Energy available through the Mechanical Impact of blowing Wind or the Wind Velocity etc. etc. and etc.

Each one of these sources has its own peculiar characteristics, its own specific potentials as well as limiting factors. Several 'trial and error' results are still undergoing microscopic considerations by the concerned experts and vested interests over the globe.

Universal consensus over the super vital issues which affect the future of entire mankind is cherished by all. But, usually for all such universal decisions the logics alone do not work. The mightiest vested interests always take the privilege to blow the last whistle.

Looking to the desperately dwindling stocks of all the existing sources of energy, many radical realignments of individual vested interests cannot be ruled out. Quite hopefully, the erstwhile fierce opponents of Wind as the primary source of energy shall be under pressure to locate or cultivate their interests in the vast field of opportunities provided by the Wind Turbines and the unlimited number of their most promising cousins and nephews.

On our planet, numerous geological and weather conditions cause the winds to blow at different atmospheric levels, at different times, in different directions, with different speeds and different momentum.

The built-up momentum of the blowing wind has the inherent force to maintain its speed and its direction. In that process, the wind exerts its forces to carry along at its own speed and direction, whatever comes in its way.

Owing to their own forces of inertia, the objects facing the invading forces of the blowing wind resist such forces with proportionately opposite reaction.

This "push-pull-and-resist" process sums up into the actual impact or the mechanical force, directly exerting on the surface coming in the way of the blowing wind. Such impact can be converted into rotary motion to drive a Wind Turbine.

The fast blowing wind is considered as the source, the carrier of the energy and not the energy in question. The energy in question is inherent in the fast blowing wind, in the shape of its physical impact.

Blowing winds are available absolutely free of cost, all the time, all over the world. Procurement of the energy from the wind does not require any digging, rigging, mining or any type of manufacturing process.

Unlike several conventional sources of energy, the primary characteristics and potentials of wind do not change at any stage of conversion of its thrust impact into mechanical form of energy. None of its elements gets consumed in any sense, degraded or dilapidated in any manner or pollute the environment.

Hence, any process of availing the wind velocity as the source of mechanical power does not necessitate any type of re-cycling or any stocks replenishments.

Wind Power is the only one which meets all the ideal qualities of the most dependable, longer lasting and affordable source of energy.

The levels of efficiency and productivity employed in any process determine the ultimate results, which can vary up to a great extent. Some fruitful possibilities are discussed in the following chapters.

IN-SIGHT

Approximately two thousand years ago, an unknown wizard in a remote village in northern part of China pioneered the concept of grinding corns etc. by using the apparent natural power of blowing wind, instead of the traditional use of animals or men.

He succeeded to channelize the blowing wind into a walled enclosure and funnel the wind to push a series of cloth masts which were mounted on the grinding stone that made it rotate on its vertical axis. In the absence of any facilities to record the history, this miraculous human achievement has all these centuries been praised through various local folk songs.

Owing to lack of commercial aptitudes, absence of personal contacts, lack of appropriate communication facilities or means of travel and transport in the most rough and tough terrains, for centuries along, this wonderful technology could not cross the boundaries of its own tiny place of birth.

Approximately around 700-AD, the painfully severe drought conditions, acute shortage of food and the question of bare survival created panic all around. In search of employment, some local workers managed to join a caravan, which was travelling from China to Persia, to sell their loads of silk and spices.

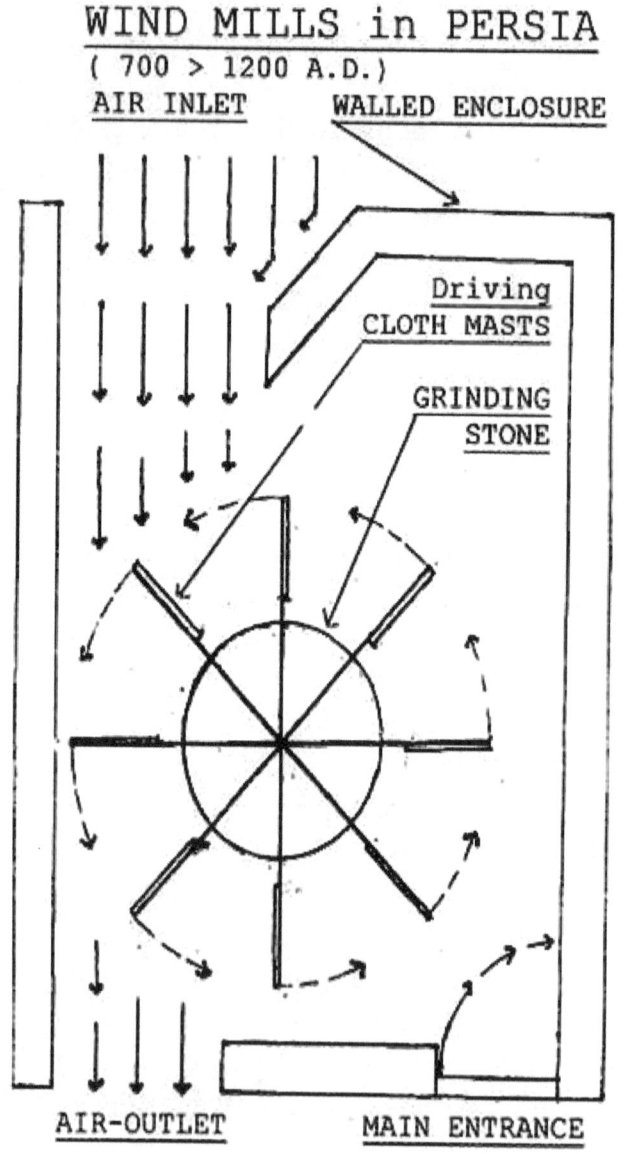

WIND MILLS in PERSIA
(700 > 1200 A.D.)

AIR INLET   WALLED ENCLOSURE

Driving
CLOTH MASTS

GRINDING
STONE

AIR-OUTLET   MAIN ENTRANCE

In Persia, those workers demonstrated their knowhow about the Wind Mills. The rulers, who also owned most of the lands there, were so deeply impressed that they decided to preserve the technology exclusively for the members of the Persian royal family.

To save the technology from going to others, they put those Chinese workers under life-time house arrest. The technology remained captive in Persia for over five hundred years. <u>Sketch-1</u> (pg.14)

Persian folk songs narrate that around 1200 AD, some Dutch youths were on excursion tours to Persia. Some of their Persian counterparts felt tempted to win-over some of their Dutch guests by exhibiting their highly secretive but boastful Wind Mills.

Thus the royal secrets mounted the wings of Cupids and travelled to join the liberties in Netherlands. Very soon, the Dutch proved their versatility and took the technology many paces ahead of Persia.

The big news of the wonderful machines became the topic of great public interest. Unlike Persia, the Dutch government and the financial institutions extended liberal support for further advancement of the technology. Enormous positive response from all channels gave birth to the legendry Dutch Wind Mills and their mushroom growth in Netherlands.
Pls. see <u>Sketch-2</u> (pg.16)

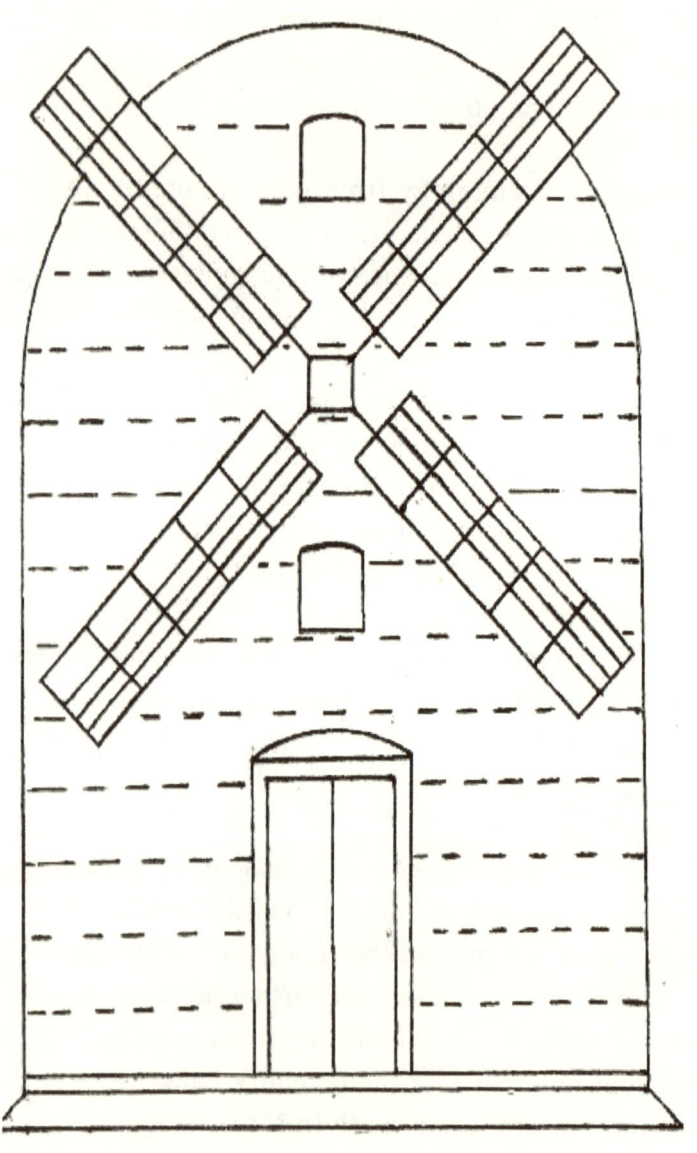

In addition to grinding corns, barley, mustard, cocoa, pepper, lime, chalk and for sawing timber and also for operating their small and medium cloth weaving units, the huge Dutch Wind Mills enabled them to manage their often flooded water drainage system.

Before the end of 18th century, they had more than nine thousand Wind Mills working around the clock all over the country. In today's world, over one thousand of them still exist as "Protected National Monuments".

Netherlands being closely connected with the western world, the Wind Mills technology spread faster and flourished all around.

Advent of Electricity in 1831 AD and the successful development of Steam Engines, affected the status of Wind Mills. Owing to lack of better comparative productivity, so many Wind Mills were closed down.

Some of the concerned engineers soon succeeded in developing the Wind Turbines with Airplane type impeller blades, mounted on the axle of an electricity generator, fitted on a high pedestal for optimum exposure to Wind Velocity.   Sketch–3 (pg.18)

Instead of grinding corns, these machines are used to produce electricity which has unlimited number of industrial or domestic applications. They installed them in the open fields, hill slopes and sea shores etc. and called them the "Wind Farms" or "Energy Parks".

In USA, the philosophy of "more the better" prevailed in every sphere. With the Wind technology still in its cradle stage, the entrepreneur started typically with the biggest possible plans, for the largest in size Wind Turbines, placed in the biggest possible Wind Parks, with biggest possible networks for the largest fields of transmission and distribution.

During 1930-40, Palmer Smith Putnam of Vermont, USA, built a 1.25 megawatt Wind Turbine with two impeller blades of stainless steel, having 175 feet rotor-span, weighing over 16 tons, rotating on horizontal axis at 28 rpm.

It was mounted on a steel structure that was taller than a twenty storey building. It was the largest ever built in the entire world. After serving for a short while, its blades broke, for reasons still not known.

Overall demand for electricity was on the high rise. The entrepreneurs in the fast industrializing world invested their major stakes in Thermal power or Hydro power projects. Wind Power was being left far behind due to lack of technology to ensure continuous electric supply when the wind velocity was not sufficient enough to turn the Wind turbines.

This provided golden opportunity for the growing Thermal power and Hydro power giants to totally negate the popularity of Wind Turbines and dislodge them from the arena.

Environmentalists had also started their worldwide agitations against the Wind Turbines. In Netherlands, they got it officially proclaimed that the Wind Energy Parks near any inhabited areas were detrimental to the public health and even on other locations they posed danger of life for the flying birds.

Exactly similar situations arose in U.K., Germany, Australia and many other parts of the world. The affected ones did not have enough of influence to challenge the one sided official proclamations.

With all the Gods on the opposite side, the ambitious Wind Turbines appeared to have gone with the wind. The still imperfect technology with limited financial resources, dwindling public and political support, crashed the ambitious projects right in the middle of the take-off runways.

At last, during 1974-81, the US Federal Government tried to revive the gusto that had been hibernating for several decades. Suddenly several plans and projects cropped up to promote the Wind Turbines.

Vertical Axis Wind Turbines, Cyclo Wind Turbines, Gyro Wind Mills, Wind Turbines with Fiber Glass Blades etc. etc. became the common topic of interest.

In 1981, the new rulers in USA brought-in new rules which crushed the worldwide process of recovery at its infant stage.

In order to accomplish any technical mission, the adequate know how and suitable equipments are the basic essentialities. But, in order to reach the goal, even the best of such combinations do not suffice. Human contributions constitute the prime driving force which ultimately determines, how, when and where the mission may or even may not reach.

In spite of several apparent merits of the Wind Power, it was officially condemned in various countries on some similar pretexts as if anyone group of opponents were exerting their might from different corners of the world. Vested interests availed the political support and thus staged their all out victory.

The Environmentalists insisted that Wind Turbines were too hazardous for the flying birds and hence it disturbed the ecological balance. Secondly, that production of electricity in the open fields near the populated areas was detrimental to the public health.

The initial technical shortcomings, such as flow of electricity while the velocity of wind was too low to drive the wind turbines, were highly orchestrated by the opposing vested interests. The Wind Power was declared as absolutely hopeless endeavor. Research and development of the available technology was officially stopped as waste of time and money. Support of the financial institutions as well as public investments evaporated from the scene. It was all for sake of the mighty vested interests.

Desperate search for the ideally dependable alternate sources of energy is still going on. At present, the virtually naked bottoms of all the known treasures of conventional sources of energy have already started the process of radical re-alignments of the prominent friends and foes in the commercial circles.

Logically, faster than sooner, the prevailing bitter realities are apt to become the blessings in disguise. The mighty trend setters and decision makers shall have to turn their necks and see the availability of more rich bounties on the other side of the horizon.

In addition to rectifying the actual technical snags, in the present situation, the promoters of Wind Power would hopefully join hands to portray the appropriate projections of the super-merits of this source of energy. That is the only way to liberate themselves from the erstwhile circumstantial impositions.

The ensuing chapters do contain some broad outlines of a few desirable technical additions and alterations in some of the existing basic concepts relating the optimum utilization of Wind Power.

SUGGESTIONS

Note :- For identity sake, suggested Wind Turbines, their Assemblies and Sub-Assemblies are referred as "Velogen" (Velocity Generators)

So far, the Wind Turbines means only the Tower type structures, supporting airplane type impeller blades, rotating on the horizontal axis, cranking a matching size electric generator which is mounted at its tail end. Pls. see <u>Sketch – 3</u> (pg. 18)

As discussed below, the Velogen Wind Turbines are designed and equipped to maintain uniform supplies of electricity under every weather condition. They also satisfy the strong demands of environmentalists all over the world. Pls. see <u>Sketch-4</u> (pg. 24)

1. Velogen TURBINE HEADS

Overall versatility of a conventional Tower type Wind Turbine increases many folds, as the turbine-head is designed with the shape of a Mushroom, a Torus or more like a fully inflated automobile tube, mounted on top of a pedestal, rotating on vertical axis.

(a) Between the upper and lower Rim of the Turbine-head, it has a number of honey-comb type air-pockets, for quick and sensitive response to any amount of wind impact, coming from any direction.

Sketch – 4   VELOGEN WIND TRURBINES

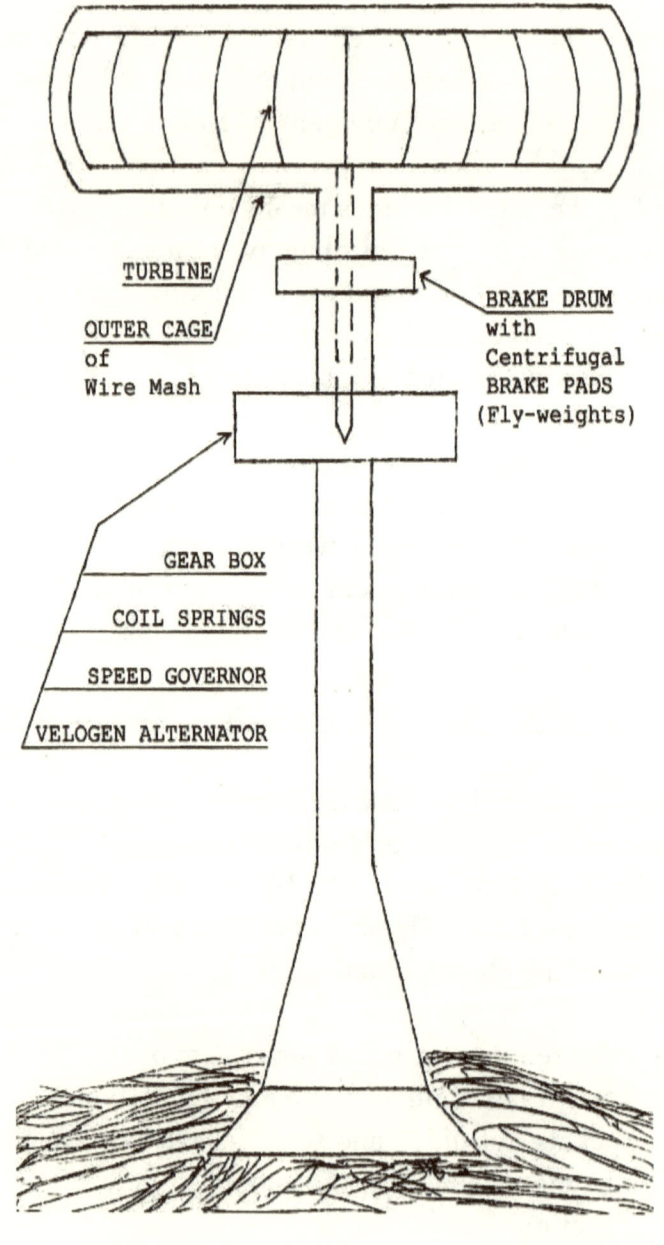

TURBINE

OUTER CAGE
of
Wire Mash

BRAKE DRUM
with
Centrifugal
BRAKE PADS
(Fly-weights)

GEAR BOX

COIL SPRINGS

SPEED GOVERNOR

VELOGEN ALTERNATOR

b) In addition to the said honey-comb air pockets, there are a number of self operating telescopic air-ports, designed to change their size and direction in quick response to the variations in wind velocity and thus increase or decrease the responsive thrust-areas in the turbine and regulate its speed even under most inclement conditions.

c) In addition to the said honey-comb air pockets and the telescopic air-ports, the Turbine Head has a number of swirling-wind-tunnels, coning-in from the wider mouth on the outer rim, towards the open corridor surrounding the axle of the turbine. It would boost the torque.

2. Velogen BIRDS SAFETY CONTROLS

The entire Turbine Head is completely covered in a suitable size cage-like formation of very strong wire-mesh-grill, to keep all sorts of flying birds etc. safely away from the turbine blades and thus guarantee that none of them can ever be sucked-in or injured or killed.

This security device does not hinder the available thrust impact of the incoming wind and at the same time it satisfies forever, the gravest concern and objections often raised by the bird lovers and the mighty environmentalist lobbies, all over the world.

3. Velogen TURBINE SPEED CONTROLS

At immediate next lower level of the encaged Turbine unit, on the main Pedestal Tower, the automatically operating Turbine Speed Control unit is housed. It comprises of a conventional Brake Drum, fixed on the body of main Pedestal and a couple of Fly-weights type Friction-Brake-Pads, mounted on the main Axle.

Incase of excessive speeding of the Turbine, the centrifugal forces generated at the Axle, push-out the Flyweights with Braking Pads, towards the inner wall of the Brake Drum and slow down the over-speeding Wind Turbine.

It keeps the maximum speed of the Turbine fully under control even under most stormy conditions .

4. Velogen POWER PLANT.

At the next lower level on the Turbine Tower, the Velogen Power Plant comprises of the following assemblies.

(A) COIL SPRINGS Assembly.

(B) AUTOMATIC GEAR BOX Assembly.

(C) A.C.ALTERNATOR Assembly.

(A) COIL SPRINGS Assembly.

It has a very high tension Strip- Coil-Spring unit, alternately switching over the connection to the main Axle of the Wind Turbine or to the A.C. Alternator, functioning through the **automatic Gear Box assembly (B), action I, II or III,** noted below.

The medium, small and miniature size Wind Turbines have inbuilt additional provisions to screw-wind the Coil Springs with the power of Wind Turbines and also by HAND-CRANKING just like winding the Clocks etc.

This provision facilitates the localized, fully independent and continuous production of electricity, even when the available Wind Velocity is not strong enough to wind up the coil spring unit through the usual Turbine Axle.

(B) AUTOMATIC GEAR BOX Assembly.

It is specially designed to automatically manage the continuous supply of driving power to the A.C. Alternator Assembly, under all weather conditions, as per functions **I, II or III** below, independent of each other.

I. When sufficient Wind Velocity
  IS AVAILABLE

  i) To route the power from the Wind Turbine
  Axle, to the Coil Spring unit and screw-wind
  the Coil Spring up to its full capacity and then
  to lodge the loaded Coil Spring unit in stand-
  by position, ready to recoil and drive the
  Alternator Disk when ever sufficient Wind
  Power **is not available**.
  ii) After completing the first stage, to divert the
  power from the Wind Turbine Axle to the
  A.C. Alternator to produce electricity.

II. When sufficient Wind Velocity
  IS NOT AVAILABLE.

  To disconnect the direct link between the
  Wind Turbine Axle and the A.C. Alternator
  Assembly and to switch over the load to the
  standing-by fully wound Coil Spring unit, to
  recoil itself and drive the Alternator for
  continuous production of electricity.

III. When sufficient Wind Velocity
  IS AVAILABLE AGAIN.

  The Gear Box will automatically repeat the
  operation **I** above, to fully rewind the Coil-
  Springs and reconnect the Alternator directly
  to the rotating Axle of the Turbine.

(C) A. C. ALTERNATOR assembly

Velogen A.C. Alternators as suggested herewith, can be considered as better choice than most of the conventional Electric Generators, for the following plus points.

1. By virtue of their much better mechanical efficiency, they need much lower power input for start-up to normal running speed.

2. Their current output and voltage can be more accurately regulated and be maintained even at lower revolutions.

3. Modified A.C. Alternators can produce variable Direct Current also, if required.

4. Unlike conventional D.C. Generators, the Velogen A.C. Alternators do not run the risk of Commutator explosion at higher speeds or any other causes of sudden break-downs.

5. Velogen A.C, Alternators need lower cost of maintenance.

The A.C. Alternators have three major components, i.e. the DISK-ROTOR,
     the STATOR, and
     the SPEED GOVERNOR,
as per details discussed below.

1. <u>Velogen Disk ROTOR</u> (<u>Sketch-5</u> (pg. 32)

Instead of conventional shapes, the suggested shape of the Velogen Disk Rotors is flat like a Compact Disk and the thinnest possible. The permanent Magnet Blocks, the surrounding Field Coils as well as the pair of Slip Rings are printed on the Disk, like Printed Circuit Boards, by using Super-Conductor printing inks.

Link from Slip Rings to the external Governor unit is maintained through spring loaded ball-points, as shown in sketch-7 & 8, page 37 & 38.

Velogen Disk has such Printed Circuits on both sides, with perfect insulation wall between them. Each side functions like a separate Disk-Rotor.

Each such unit of Two-in-one Disk Rotors revolves within the flux field of two separate Stators of suitable size and shape, sandwiching each Rotor Disk separately. Hence, for one revolution of the disk, the two-in-one Rotors unit produces double the amount of electricity. Pls. see <u>Sketch – 7</u> (Pg. 37).

The two-in-one Disk Rotor is mounted on a Double Ended Axle with both its ends having 'tapered-pin-point' shape, resting/floating on a hard jewel or industrial diamond at each end.

Such Driving Axle design minimizes the rotating friction and thus requires minimum of driving power for the Rotor and enhances the working mechanical efficiency of the Rotor.

Axle of the two-in-one Disk Rotor is driven by a spindle with fine helical grooves, powered through either one of the following sources. i.e.

Wind Turbine > Coil Springs > Governor > Disk Rotor
(when sufficient wind velocity **is available**)
OR
Coil Springs > Governor > Disk Rotor
(when sufficient wind velocity is **not available**).

Pls. see Sketch – 5 (page 32)

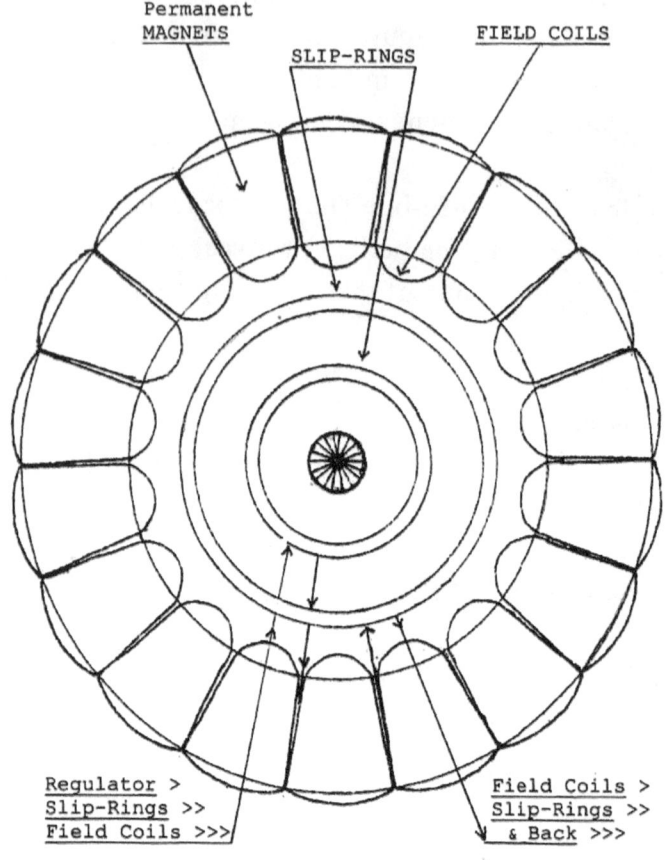

Permanent
MAGNETS

SLIP-RINGS

FIELD COILS

Regulator >
Slip-Rings >>
Field Coils >>>

Field Coils >
Slip-Rings >>
& Back >>>

<u>Sketch - 5</u>

<u>Velogen Disk ROTOR</u>

2. Velogen Disk STATOR

The Velogen Disk Stator is also of the shape and size corresponding to the Velogen Disk Rotor.

The Core Segments as well as the Collector Coils are also printed in Super Conductor Ink, like the Permanent Magnets on the Disk Rotor.

Each Double sided Disk Rotor is "sandwiched" between two Velogen Disk Stators, one on each side. For each revolution of the Rotor, it will produce double the amount of electricity.

Pls. see Sketch – 6 (pg. 34)

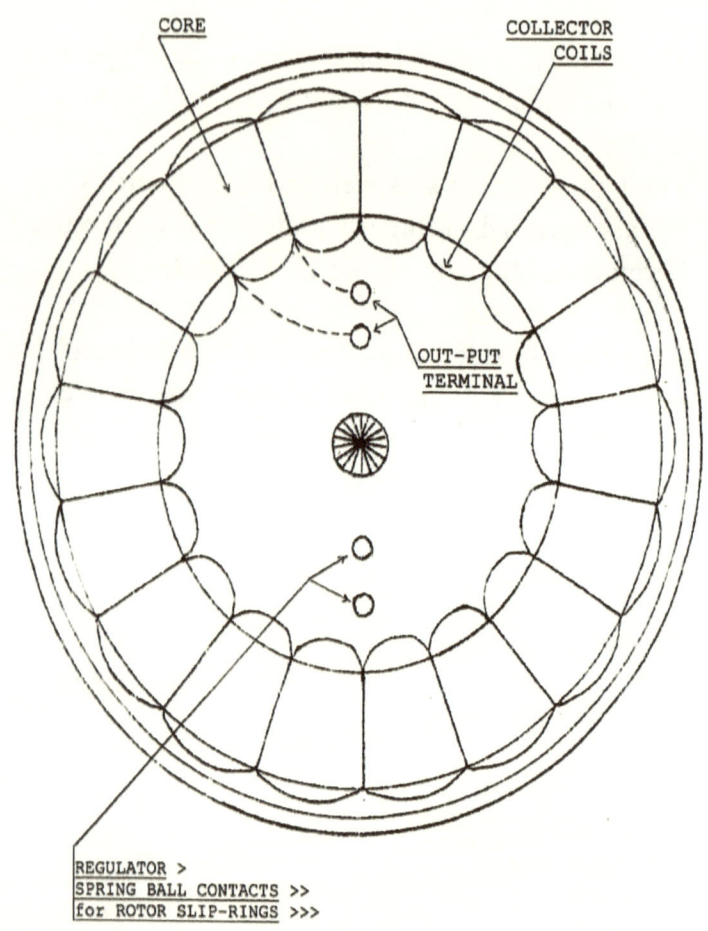

CORE

COLLECTOR
COILS

OUT–PUT
TERMINAL

REGULATOR >
SPRING BALL CONTACTS >>
for ROTOR SLIP–RINGS >>>

Sketch – 6

Velogen Disk STATOR

34

3. SPEED GOVERNOR

Its prime function is to 'fine-tune' the end speed of the Disk Rotor within the A.C. Alternator.

It draws its in-put power either directly from the Wind Turbine main Axle or from the recoiling Coil Springs.

This switch-over is carried out automatically by the Gear Box Assembly, according to the Wind Velocity situation.

It operates just like conventional governors as provided in the age old grand-pa's gramophone.

Its main in-put drive spins a number of small flyweights which are mounted on steel strips with their one end fixed on a small flywheel which slides on a turret base, responding to the dynamic position of the spinning flyweights, varying according to the speed.

An adjustable soft brake pad is positioned to apply on the wall of the flywheel to govern its speed and thus, the speed of the out-put spindle.

The output spindle drives the Velogen Disk Rotor at the desired speed by setting the position of soft brake pad on the flywheel.

Actual size, shape or even the functional details of the Velogen Speed Governor may vary from model to model.

The One in the giant size Tower type Velogen Wind Turbine is of a much more rigid and robust design and operates differently from the one in a medium, small or a miniature size.

But in each case they do perform the most significant role of ensuring fully controlled uniform speed of the Velogen Disk Rotors and thus their smooth output.

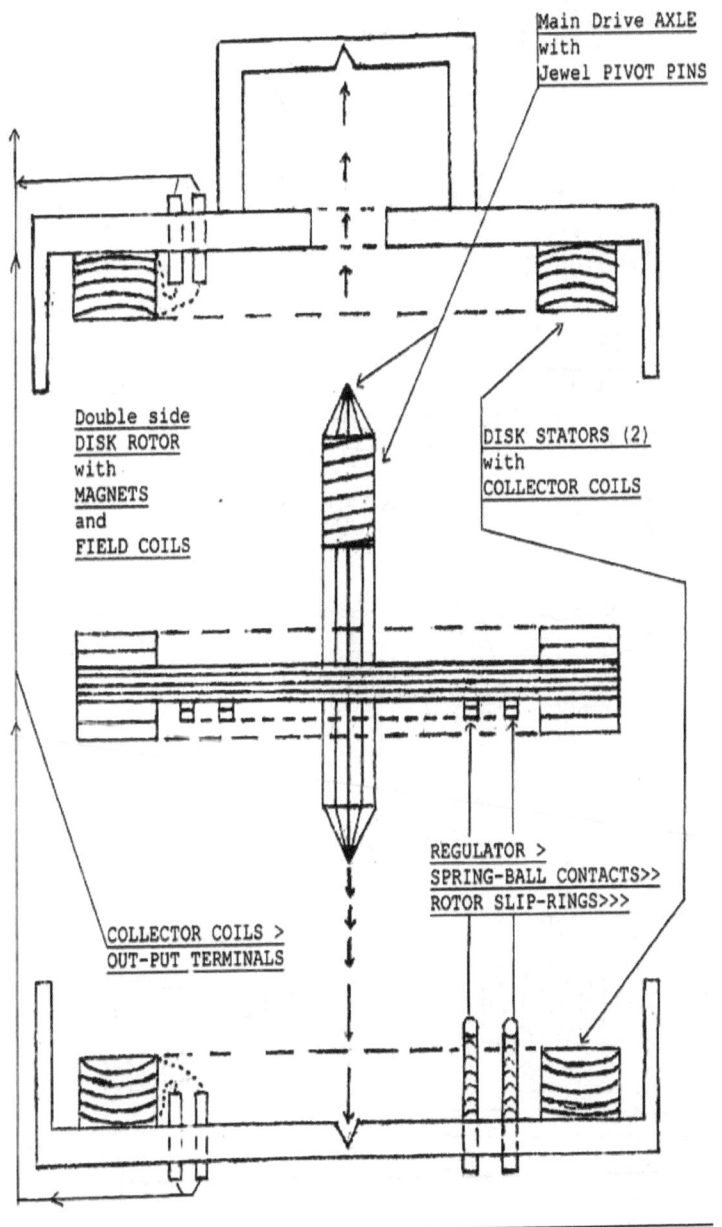

Main Drive AXLE
with
Jewel PIVOT PINS

Double side
DISK ROTOR
with
MAGNETS
and
FIELD COILS

DISK STATORS (2)
with
COLLECTOR COILS

REGULATOR >
SPRING-BALL CONTACTS>>
ROTOR SLIP-RINGS>>>

COLLECTOR COILS >
OUT-PUT TERMINALS

SKETCH – 7   Velogen A.C. ALTERNATOR

SKETCH - 8

Velogen ALTERNATOR

Disk STATORS (2)
with CORE &
COLLECTOR COILS.

Disk ROTORS (2)
with Double side
MAGNETS &
FIELD COILS.

Main Drive AXLE
with
Helical Grooves &
JEWEL PIVOT PINS.

REGULATOR >
ROTOR SLIP-RINGS>
FIELD COILS & Bac

COLLECTOR-Coils >
OUTPUT TERMINALS.

OPORTUNITIES

In view of the suggested changes in shapes, sizes or other functional details of the traditional Wind Mills and Wind Turbines, the horizons keep opening up further and further beyond. Numerous unimagined and unseen opportunities come to light.

Scope for the Velogen Wind Turbines is apparently as vast as any flight of imagination can care or dare to reach. Probable launching pads are discussed below.

1. Velogen AUTOMOBILES

While designing the size and shape of the body of any automobile, the opposing wind is always considered as a negative force. All the skills for designing the external contours of the automobiles are invested in trying to dissipate or to minimize the direct impact of the opposing wind, while the automobile would be in motion.

Negative force of the opposing wind, if properly channelized through the front grill of any type or size of an automobile, it can conveniently be funneled into the sizeable space available under the bonnet. This space can be fruitfully utilized to house a number of suitable size Wind Turbines operating with the impact of incoming air streams and thus produce electricity.

Ultimate results in the process of generating electricity on board any vehicle bank upon the following factors.

a) Design and shape of the Front Grill and its further contours to funnel the accumulated wind streams into the bonnet chamber.

b) Inside the bonnet chamber, the design as well as the positioning of various functional layouts to manage the direction as well as the intensity of the swirling impact of the incoming wind streams.

c) Size, quantity and the actual placement of the suitably updated ultra feather touch Velogen Wind Turbines and the Alternators. (Pls. see sketches 5, 6, 7 & 8, pages 32, 34, 37 & 38 and relevant detailed discussions)

   Any single unit of the largest possible Wind Turbine with Alternator shall require larger amount of wind impact to reach the productive speed. It will fail to produce electricity at slow vehicle speeds with slow wind streams.

   Hence, instead of "Big is Best" philosophy, it will be better to deploy the largest effective number of technically efficient Velogen Wind Turbines, possibly in a cascading formation, to obtain maximum output with any input.

Suitable size Car Batteries can provide the start-up motion to the vehicle. As the speed increases, the air gushing-in through the front grill would rotate the Wind Turbines to produce electricity.

Even at the city running speeds, available thrust of the incoming swirling wind would be sufficient to operate the feather touch Wind Turbines in their cascading formation. Cumulative quantum of electricity so produced would be sufficient enough to keep the Batteries always fully recharged as well as provide power to operate the Traction Motors and thus to keep the vehicle absolutely self sufficient even at city running conditions.

Thus, whether in the city or on the highways, the Velogen powered automobiles shall never need to recharge their batteries through any external plug-in source. They shall never require any type of fuels or 'fill-ups' to drive for any number of miles.

The moments are close-bye, when the old common joke about the future automobiles doing a million miles per gallon would become a living reality for all times to come.

Of course, it will require a lot of detailed scientific research and design development, a lot of dedicated team work and a big lot of concerted efforts in quest of perfection.

2. Velogen RAILWAY TRAINS

Second most fruitful application of the Velogen Wind Turbines would be on the Railway Trains, Trams and Trolleys, all over the world.

Adequately designed suitable size Velogen Wind Turbines, housed on the roof tops of each one of the Railway bogies and also at all suitable spaces within all of their chassis frames, would always face enough of Wind velocity to drive any large number of suitable capacity Alternators.

The abundant availability of extremely powerful wind blasts on the roofs as well as in between the chassis of a running train can be fully dependable source of energy to generate enough of electricity which can suffice for the lighting, utilities as well as the main traction load of any size of railway train.

A number of suitably interlinked large capacity Wind Turbines fitted on top of the wagons and the available spaces on the chassis, shall have the capacity as large as the Locomotives.

In addition to making each railway train fully self-sufficient in fuel inputs, the 'carry-along' power houses will save enormous amount of running costs, maintenance as well as overhead expenses. Above all, it would be absolutely pollution free.

3. Velogen SHIPS & VESSELS

Richest treasure of the natural Wind velocity is available all around the year on vast sea shores, water-ways and open seas. The decks, the bridges and numerous other spots on the sailing vessels, ships and tankers, container carriers, cruise boats etc. etc. get plenty of wind velocity for most of the time. It can be sufficient enough to keep large number of Wind Turbines always producing large flow of electricity on board. Proper planning can provide sufficient supply of electricity for lighting and other utilities as well as all the power required to drive Traction Motors of most of the vessels.

Old and discarded vessels of any size and shape can be suitably equipped to perform as absolutely self sufficient "Floating Power Houses" for supplying electricity to any on-shore vicinities or any off-shore projects or establishments, without any external inputs.

Since all of these units always function without smoke, without any noise and absolutely without any emissions, these floating power houses can be camouflaged and be safely deployed for Defense requirements.

The total cost of operating would obviously be, only the actual cost of lubricants and the actual cost of periodical maintenance.

4. Velogen WIND FARMS.

Tower type large Wind Turbines (sketch-4, pg. 24) should be planted on all the convenient sea-shores, river-banks, hill-slopes, valleys, plains, fields and all such locations where ever natural wind velocity is available for any justifiable time.

Instead of continuing with the long distance transmission lines, a larger number of localized small pockets of Wind Farms or Energy Parks only with locally viable total number of Tower type Wind Turbines should be installed and interlinked through suitable network of Grids and Loops, preferably to serve the bulk consumers in any nearby vicinities.

It will save the initial cost of extra long distance transmission lines and other additional apparatus, plus the recurring costs of maintenance and huge overheads.

5. Velogen REMOTE INSTALLATIONS.

By installing suitable size and numbers of Velogen Wind Turbines, each one of the on-shore, off-shore or high seas Drilling Riggs or any other such installations in every remote areas shall become absolutely self sufficient for all of their needs of electricity for lighting, utilities as well as for all of their equipments.

All such "in-house" Power Houses can practically serve all their needs without any fuels or other recurring in-put. On top of that, it would be absolutely pollution free.

6. Velogen MULTI-INPUT WIND TURBINES

Medium and small size Velogen Turbines would have some additional features in order to enhance their versatility and substantially enlarge their fields of application.

a) In addition to the usual Velogen Wind Turbine, there shall be provision to drive the Alternator through an attachable or extendable or flexible-cord type axle-shaft, operated by 'rowing-fins' type rotary blades, to avail the hydro power of any water stream.

b) The medium, small and miniature size Velogen Wind Turbines would have inbuilt additional provisions to screw-wind the Recoiling Coil Springs with the rotary power of the Wind Turbines and also by HAND-CRANKING , just like winding the old Time Pieces and Clocks. .

Both of these provisions would facilitate continuous production of electricity, without depending on Wind Power. Such multi-input Turbines shall be the most useful in remote areas, dense forests or for defense purposes as well as for domestic or any emergency needs.

7. Velogen "GENIES"

In almost all the remote areas where electricity could not reach so far; where people still live in the past medieval era, the medium and small size Velogen Wind Turbines, with the multi-input additional features as discussed above, fixed on any bench or any portable pedestals, would surely perform several unseen magical feats, just like the proverbial "Genies".

These modern GENIES would serve to ....
i) Light up their huts, barns and their paths.
ii) Energize their cooking stoves and ovens.
iii) Supply power for their heaters and coolers.
iv) Fetch water to their own door steps.
v) Irrigate their small crops and cultivations.
vi) Operate their farm equipments, cutters, thrashers etc. for better agricultural yields.
vii) Enlighten their schools, libraries and community centers.
viii) Equip their dispensaries and health centers.
ix) Promote cottage industries, handicrafts and artisan shops.
x) Promote small food processing, preservation and marketing units.
xi) Create lots of self-employment and other job opportunities.
xii) Enhance overall human productivity and all round prosperity.
xiii) Keep them "related" to the "other world".

God knows whether it could be called the human evolution or a part of universal industrialization.

8. Velogen **DOMESTIC POWERHOUSES**

The cities and suburbs also hold equally bright scope for the multi-input medium and small size portable as well as the fixed units, looking like table fans and pedestal fans.

It will enable every household to own their exclusive Electric Power House, operating noiselessly, without smoke and also without any pollution, sitting pretty in any of their windows, balconies, courtyards, backyards, lawns, driveways, or even on the roof-tops.

All such Velogen Wind Turbines would be able to operate with Wind power as well as with Hand Winding Recoiling Coil Spring assemblies. They shall be able to serve each individual household for entire requirement of electricity, free of cost.

All that one has to do is, to place the unit in any part of the house where there may be more chances of facing the wind velocity. Keep the Coil Springs fully hand-wound and in "stand-by" position for instant response, as and when required. Plug-in the Velogen 'out-put' cord into any of the electric wall-sockets.

As the need be, each household can plug-in more than one such Wind Turbine unit at a time.

Each unit would not cost more than an average air conditioner. Lifetime total cost shall be, the initial cost plus cost of periodical cleaning and lubrication. There shall not be any bills to pay for electricity.

9. Velogen for EMERGENCY SERVICES

In addition to the handy and versatile domestic Power Houses, there would also be some custom-built models to suite the specific requirements of the hospitals, dispensaries, schools, libraries, community centers, street lights and road-side public utilities etc.

As there is no human control over any of the natural calamities and disruptive weather conditions, come what may, the multi-input Velogen Wind Turbines shall keep all the essential services running smoothly as desired. Of course they would require time to time hand winding the Coils unit.

10. Velogen MINIATURE POWERHOUSES

There is enormous scope for the miniature size Velogen Turbines, driven by the Wind Power or by the inbuilt facility to operate them through Hand Wound Recoiling Coil Springs.

The Emergency Relief Units, Geological and other Surveyors, Rangers, Border Patrols, Security Guards, Miners, Mountaineers, Hikers, Hunters, Campers and many more would benefit from such tiny, handy and versatile gadgets.

Velogen Wind Turbines with only six inches diameter, when mounted on top of a tent in any snow-clad area, or anywhere indoors, it will light a lamp, warm-up the bed and make the entire tent cozy and comfortable.

While facing the desert like burning hot weather conditions, a portable Wind Turbine, even when running on the power of recoiling hand wound coil springs, can produce enough of electricity to operate a small desert cooler in the tent and also keep the food safe and drinking water cool.

A tiny Velogen Wind Turbine of less than three inches diameter, if worn like a peak-cap, or a band around the bicep or conveniently tucked into any button hole, can keep the interior of a snow-suit warm enough even in the most frigid polar areas or any other freezing cold zones.

Major achievement of these Wind Turbines will be, to localize the production, transmission, distribution as well as the consumption of electricity.

Looking at the utility, the necessity and also the apparent marketing potential, it would be more appropriate to produce at least ten thousand units of multi-input medium and small size Velogen Wind Turbines and another twenty thousand units of the miniature size, for every ten units of the giant size Tower type Velogen Wind Turbines.

With these great achievements, the high initial costs of the conventional layouts of Transmission lines, Grids, Transformers and other Distribution networks, as well as their most expensive maintenance shall no more remain essential.

The modern Personal Power Houses shall virtually liberate the consumers of electricity all over the world, from their recurring liabilities towards the unlimited costs of conventional management, the huge amounts which are being regularly borne by the consumers and consumed by the overheads, leaving behind some simply complex statistics for public consumption.

Such big steps taken with full determination, will most obviously take full care of the global warming dilemma and environment pollutions, on permanent basis.

Wind Power is obviously the one and the only one, " MOST ENERGETIC SOURCE OF ENERGY".

GO-AHEAD *my* READERS

Go ahead with the clearer and brighter dreams about the on-going present as well as the future tense. Go ahead with the non-stop crystallization of your views, your ambitions, your plans and your projects. Go ahead with the expedition towards perfection, towards the successful accomplishment of your mission.

Based upon the conceptual sketches of the desirable launching pads, there is still a very long way to go. Above all, the concepts are mere concepts. The inherent and integral human factor cannot be ignored or over ruled. They always necessitate advanced intellectual support and more appropriate scientific approval. In this process, there is absolutely nothing to feel shy about.

Be realistic, be systematic and remain persevering. Always think high, plan high and still remain in touch with your foundations below the ground level.

Success is no one's monopoly. Any-one, including some-one, may be the next-one.

So, GO-AHEAD, as ahead as you can.

humbly,   G. P. Walia

Notes & Remarks

Notes & Remarks

<u>Notes & Remarks</u>

www.ingramcontent.com/pod-product-compliance
Lightning Source LLC
Chambersburg PA
CBHW021925170526
45157CB00005B/2185